MICROSCOPY HANDBOOKS 03

Specimen preparation for transmission electron microscopy of materials

T0174115

Peter J. Goodhew

Department of Materials Science and Engineering, University of Surrey

Oxford University Press · Royal Microscopical Society · 1984

Oxford University Press, Walton Street, Oxford OX2 6DP

London New York Toronto
Delhi Bombay Calcutta Madras Karachi
Kuala Lumpur Singapore Hong Kong Tokyo
Nairobi Dar es Salaam Cape Town
Melbourne Auckland

and associated companies in
Beirut Berlin Ibadan Mexico City Nicosia

Royal Microscopical Society
37/38 St. Clements
Oxford OX4 1AJ

Published in the United States
by Oxford University Press, New York

Transferred to Digital Printing 2006

British Library Cataloguing in Publication Data
Goodhew, Peter
Specimen preparation for transmission
electron microscopy of materials.—
(Microscopy handbooks ; 3)
1. Electron microscope, Transmission
2. Materials—Testing
I. Title II. Royal Microscopical Society
III. Series
620.1'1299 TA417.23
ISBN 0−19−856403−1

Publisher's Note
The publisher has gone to great lengths to ensure the quality of this
reprint but points out that some imperfections in the
original may be apparent

Contents

1   **The specimen: what we are trying to achieve**   1

   1.1   Thinking ahead   1

   1.2   The ideal specimen   2

2   **Initial preparation of sheet or disc**   5

   2.1   Cutting the slice   5

   2.2   Preparing flat faces   6

   2.3   Chemical thinning solutions   7

   2.4   Making a disc   10

3   **Final thinning**   12

   3.1   Electropolishing   12

   3.2   Thinning a sheet: the window technique   14

   3.3   Thinning a disc: the automatic jet polisher   16

   3.4   Recipes for electropolishing   17

   3.5   Ion beam thinning   21

   3.6   Other final thinning techniques   24

   3.7   Thinning specific regions of a sample   26

   3.8   Thin foils from just beneath the surface   26

   3.9   Cross-sections perpendicular to the original surface   27

4   **Replicas**   29

   4.1   The role of the replica in modern microscopy   29

   4.2   The extraction replica   30

5   **Mounting and storing specimens**   33

   5.1   Grids   33

   5.2   Disc specimens   33

   5.3   Thinned foils   33

5.4   Support films   34

5.5   Specimen storage   35

Appendix. Materials and suppliers   37

References   39

Index   41

The specimen: what we are trying to achieve

Electron microscopy is a powerful and fascinating tool for the investigation of structure and composition on a fine scale. Considerable skill is required to operate the microscope itself, to take high quality micrographs, and to interpret the resultant images. These topics are dealt with in companion handbooks. In this volume we shall be concentrating on the essential practical steps which must precede the microscopy itself and which are absolutely crucial to successful and meaningful use of a transmission electron microscope. Without a thin undamaged specimen even the most skilful microscopist is helpless.

Let us first state the primary objective of specimen preparation. It is to prepare and mount successfully in the microscope a thin specimen from which it will be possible to deduce accurately the structure, composition, and often also the behaviour of a larger sample of the material. There are several possible general approaches to achieving this objective. The most common is to thin a large piece of the material, for example by electropolishing, or to disperse it into pieces which are small enough to be transparent to electrons. This is the subject of Chapters 2 and 3. However, two alternative approaches are useful in a limited number of cases. It is sometimes possible to prepare the material directly in the form of a thin sheet, for example by vacuum deposition. Indeed, this often produces a marvellous specimen in all respects except the most important − it may not be representative of the material on a larger scale. However, if the object is to study thin deposited films then this simple technique is clearly highly appropriate. A third approach to specimen preparation is to extract from a larger specimen only certain components of its structure for study in isolation. This can be convenient but of course cannot give a total picture of the microstructure. The major technique of this type is carbon extraction replication, which is dealt with in Chapter 4.

1.1. Thinking ahead

Before deciding on a specimen preparation route (that is, if it is not already too late, before going straight to Chapter 3) it is wise to consider what transmission electron microscopy (TEM) techniques are to be used on the specimen. This will help to decide what are going to be the most important attributes of the final foil. For example, if large-scale low magnification information about the structure is required, uniformity of thinning and large size of thin area are paramount. On the other hand for weak-beam microscopy involving long exposures specimen stability is very important. Again, if the specimen is to be analysed (by X-ray or energy loss techniques) it will be necessary to avoid segregation or preferential leaching of particular elements in the specimen. A further point is that if the electron beam is to be kept stationary on a very small region of the specimen (for microdiffraction, electron energy loss spectroscopy, or convergent beam diffraction) the specimen will need to be as free as possible from mobile surface contaminants. It is not easy to meet all these criteria so it is as well to determine in advance which is most important for a particular application.

A second way of thinking ahead is to assess the likely effect on the specimen of the various possible preparation procedures. This must be considered at all stages of preparation, and not just in the final stages when the specimen is obviously thin and delicate. Finally, it is clearly necessary to bear in mind what facilities and equipment are available locally. It is unlikely that all the equipment described in this booklet will be present in the laboratory. However, it can be demonstrated that a relatively modest outlay on reliable specimen preparation equipment is a good way of maximizing the effectiveness of the transmission electron microscope in every project. Most of the equipment described in this monograph costs less than 2 or 3 per cent of the value of the microscope it is intended to support.

1.2. The ideal specimen

The attributes of the perfect TEM specimen are listed in Table 1. Of course, these are never all realized simultaneously in practice; however, it is helpful to know what is being aimed at.

Table 1

The ideal TEM specimen is	Representative
	Thin
	Stable
	Clean
	Flat
	Parallel-sided
	Easily handled
	Conductive
	Free from segregation
	Self-supporting

Let us consider these attributes in turn. 'Representative' appears at the head of the list because without this all the other items are worthless. The specimen must accurately reflect the nature of the bulk material. We must be alive to the many possible ways in which the tiny area we finally photograph may not be typical. Among the more obvious problems are the introduction of dislocations or point defects due to mechanical damage during preparation, the loss of particular phases because of differential thinning, the unwitting selection of thin areas from only one phase in a two-phase material, or the non-random occurrence of particular crystal orientations in a thin polycrystalline foil. The most easily avoidable of these problems, involving damage during preparation, are considered in the next two chapters as each technique is introduced. The more subtle problems of non-representativeness must be left to the intelligence and materials experience of the microscopist.

It seems obvious that a TEM specimen must be 'thin', but it proves rather difficult to quantify this statement. The concept of 'thin' must depend on the material (the electron penetration decreases as atomic number rises), on the accelerating voltage V available (useful penetration rises with V), on the imaging resolution required (high resolution requires very thin specimens), on the sideways spread of the electron beam that can be tolerated (in attempting for example to analyse a small region), on the size of second-phase particles which are required to be included within the foil, or on whether Kikuchi lines are required in the diffraction pattern. All that can safely be said is that the ideal thickness will probably lie between 100 Å (for lattice resolution) and several micrometres (for the high voltage electron microscopy (HVEM) of light elements). In practice most specimen preparation techniques lead to foils which taper from very thin to too thick, and the appropriate region for the type of image or analysis in hand must be chosen.

There are two aspects of specimen stability which are of concern. The response of the specimen to the electron beam should not lead to either chemical change or specimen drift. At the same time the ideal specimen should survive unchanged in air at room temperature for many years – it can then be re-examined at a later date. There is not a great deal which can be done about the latter – if mild steel is studied it will oxidize in damp air. The only solution is to keep prepared specimens in an appropriate atmosphere – in the case of mild steel in a desiccator in a refrigerator. Specimen stability in the microscope can be improved if the thin region of the specimen is supported by a much thicker region, as generally occurs with a disc specimen.

The cleanliness of a specimen is obviously of great importance since artefacts arising from dirty polishing solutions or simple laboratory dust are irksome and unsightly when superimposed on interesting microstructures. However, less obvious but more important are hydrocarbon layers which can give rise to contamination of beam-irradiated areas. Prevention of this effect remains one of the most difficult problems in TEM.

It is preferable for a specimen to be flat and parallel-sided for a variety of reasons. This will, in a crystalline specimen, eliminate thickness fringes and reduce the number of extinction contours. Also, if either the composition or the microstructure of the specimen is to be analysed quantitatively it is necessary to assume that the foil is of uniform thickness in the region photographed in order to arrive at the volume of the region. A similar consideration dictates that the specimen should contain no segregation of particular species to its free surfaces.

Finally, the specimen must be placed in and removed from the microscope without damage, and this is most easily achieved if it is self-supporting and can be handled with tweezers. If a support grid must be used then it will inevitably obscure some areas of the specimen (which, by Murphy's Law, will be the best thin areas).

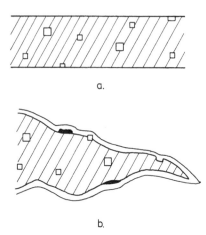

Fig. 1. (a) An ideal specimen which is flat, parallel-sided and undistorted. (b) A more usual specimen which tapers, droops and is covered with a layer of oxide, contaminant, and dirt. Some of its second-phase particles have been etched out and some stand proud of the foil.

If the specimen is electrically and thermally conductive then both charge build-up and too great a rise in specimen temperature can be avoided during examination.

It takes no clairvoyant to appreciate, before reading the rest of this monograph, that a specimen which fulfils all the criteria discussed above is only rarely produced. However, an appreciation of the ideal attributes of a specimen, and of the reasons why they represent an ideal, should help in the selection of techniques from the next three chapters which will be appropriate to the material under investigation and the reasons for its examination.

As a brief summary of this section a comparison of the 'ideal' and the 'typical' specimen is shown in Fig. 1. This is not intended to be entirely flippant!

Initial preparation of
sheet or disc

The thinning of bulk material to electron transparency is generally achieved in two stages, which we shall call initial preparation and final thinning. Most of the final thinning techniques detailed in Chapter 3 work best on thin sheet or 3 mm diameter discs with flat or polished surfaces. In order to minimize the time taken by final thinning and to maximize the chances of ending up with a large parallel-sided thin area the initial preparation procedure should produce a sheet or disc about 0.1 mm (100 μm) in thickness. The lower limit on this thickness is set by the nature of the specimen material and its susceptibility to damage during preparation. A very hard material could usefully be prepared to a thickness of less than 50 μm, whereas a ductile single crystal will be difficult to thin below 200 μm without damage.

We shall consider initial preparation in three steps, illustrated in Table 2, although it is sometimes possible to omit some of these.

Table 2. *Common techniques for initial preparation*

(1) Cut slice	(2) Smooth faces	(3) Make disc
Saw	Hand polish	Punch
Acid saw	Mill	Spark machine
Wire slicer	Lap	Hollow drill
Slitting wheel	Chemical polish	Ultrasonic drill
Spark machine		Cut from rod
Cleave		
Microtome		

2.1. Cutting the slice

The first step is to cut a rough slice from the bulk specimen, bearing in mind when choosing the section that the final viewing direction will be perpendicular to the slice. At this stage the slice will probably have two rough surfaces and the thickness of the slice must be determined by the likely depth of damage caused by the chosen cutting technique. A conventional hacksaw, even with fine teeth, will probably do most damage and may easily affect the structure as far as 1 mm from the cut in a soft metal. Three less damaging techniques are the spark machine (used with a low energy spark), the diamond slitting wheel, and the rotating wire saw which carries an abrasive slurry through the specimen. The choice of technique depends on the material. Spark machining can only be applied to conductors and tends to be slow.

A slitting wheel can be much faster and is useful for non-conductors, but there is a risk of overheating the specimen. The wire slicer is a delicate device, widely used for cutting single crystals, but it imparts very little damage and progress can be monitored very easily. The least damaging, but most difficult, technique is the acid saw, in which a moving string carries acid (or some other appropriate solvent) through the cut. This is very gentle since the action is essentially chemical rather than abrasive, but it is consequently rather difficult to control and the preparation of a thin slice is not easy.

The four techniques are illustrated schematically in Figure 2. These devices are all available commercially in small sizes suitable for TEM preparation. A list of manufacturers and suppliers is given in the Appendix.

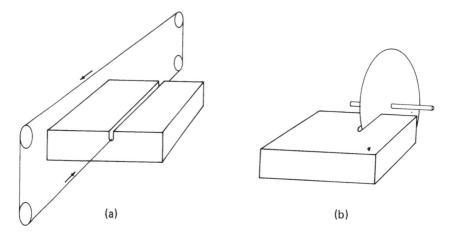

(a)   (b)

Fig. 2. (a) The acid saw, wire saw, and spark machine all employ a moving thread. In an acid saw this is a string which passes through an acid bath. In a wire saw it is a wire passing through an abrasive slurry. In a spark machine the whole device is immersed in paraffin and a spark is established between the moving wire and the workpiece. (b) A slitting wheel: the diamond-impregnated wheel passes through a slurry of lubricant and/or abrasive.

2.2. Preparing flat faces

Having cut a slice of thickness between 3 and 0.5 mm it is now necessary to prepare its faces as flat and parallel-sided as possible. The best way to achieve parallel faces is by machine milling or lapping, and the latter is preferable, if available, since the depth of damage can be reduced to quite a small value by using a fine abrasive. Certainly parallel-sided sheets of $100\,\mu m$ thickness or less can be produced from most materials by lapping with a 600 grit abrasive powder. Mechanical polishing by hand is a cheap and simple alternative to machine methods. If only a small sample is to be thinned (e.g. a 3 mm disc), then a very simple jig can be employed to control the thinning. Such a jig is shown in Fig. 3: it can be bought (see Appendix) or made in the workshop. With such a device it is possible to thin a disc to less than $50\,\mu m$ while maintaining highly parallel faces. The desired final thickness

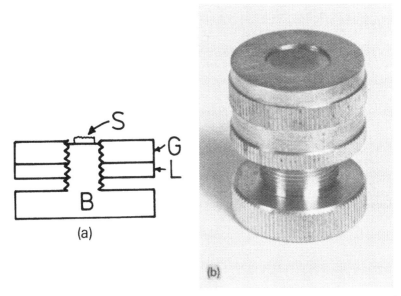

Fig. 3. A simple jig for hand polishing. The specimen (S) is glued to the central post (B) while a guard ring (G) is set to protrude by an amount equal to the desired final thickness. This is held in place by a lock ring (L).

can be set using the lock rings, and the thinning can then be done on conventional metallographic papers and wheels.

A major problem with all the mechanical techniques for producing smooth faces is to secure the slice to a substrate so that it can be handled. Each of the techniques described above can only be applied to one surface at a time, and the usual approach is to polish or lap one face and then to turn the slice over and re-fix it to the substrate with as thin a layer of adhesive as possible before finishing the second surface parallel to the first. Many adhesives are suitable but two of the most useful are a low-melting-point wax such as dental wax or a solution of polymethylmethacrylate (PMMA; Perspex) in ethyl acetate. In extreme cases double-sided adhesive tape can be used, but this is generally too thick and too soft for ideal security. Whatever adhesive is used, care must be taken to wash it off thoroughly or subsequent final thinning is likely to be uneven.

2.3. Chemical thinning solutions

The least damaging method of thinning a slice is by chemical polishing. This of course removes damage introduced by any previous mechanical stage but it is difficult to keep the faces of the slice perfectly parallel. Although chemical thinning machines have been built (see for instance Goodhew 1972) it is generally possible to polish both faces of a slice simultaneously in a simple beaker of thinning solution. If plenty of material is available, it is even possible just to immerse a slice in the solution without any attempt to prevent preferential attack at its edges. The resultant slice is rather smaller but if the faces are acceptably polished this is not normally of great consequence. The key to successful chemical thinning is the correct choice of thinning solution. These solutions generally contain an acid, with if necessary the addition of a solvent for any solid oxides which may be formed. Table 3 gives a list of solutions which have been used successfully in the past. These should be considered as starting suggestions; it may be necessary to experiment with temperature or concentration in order to find the ideal for the material under investigation. Detailed references to the solutions listed in the

Table 3. *Chemical thinning solutions*

Material		Solution (use at room temperature unless otherwise stated)
Many metals		20% HNO_3, 80% methanol
Ag & alloys		50% HNO_3, 50% H_2O
Al & alloys	1	40% HCl, 60% H_2O + 5 gl^{-1} nickel chloride
	2	200 gl^{-1} NaOH in H_2O, 70°C
	F3	64% H_3PO_4, 18% HNO_3, 18% H_2SO_4, 80°C
	4	50% HCl, 50% H_2O + a few drops H_2O_2
	5	94% H_3PO_4, 6% HNO_3
Be	F	75% H_3PO_4, 7% chromic acid, 5% H_2SO_4, 13% H_2O, 60°C, slow.
Cd	1	20% HNO_3, 80% methanol
	F2	40% H_3PO_4, 60% H_2O
	F3	12 g KCN in 100 ml H_2O + 2 g Cd(OH)$_2$
Co–Fe		50% H_3PO_4, 50% H_2O_2
Cu & alloys	1	80% HNO_3, 20% H_2O
	2	50% HNO_3, 25% CH_3COOH, 25% H_3PO_4
	3	40% HNO_3, 10% HCl, 50% H_3PO_4
	4	60% CH_3COOH, 30% HNO_3, 10% HCl
Fe & steels	1	30% HNO_3, 15% HCl, 10% HF, 45% H_2O, hot
	2	35% HNO_3, 65% H_2O
	3	60% H_3PO_4, 40% H_2O_2, and other mixtures
	4	33% HNO_3, 33% CH_3COOH, 34% H_2O, 60°C
	5	34% HNO_3, 17% CH_3COOH, 32% H_2O_2, 17% H_2O; add H_2O_2 just before use
	6	40% HNO_3, 10% HF, 50% H_2O
	7	5% H_2SO_4 saturated with oxalic acid, 45% H_2O, 50% H_2O_2; add H_2O_2 just before use
	8	95% H_2O_2, 5% HF
GaAs, GaP	F1	1–4% Br in methanol
	F2	50% HCl, 50% HNO_3
	F 6	42% HNO_3, 14% HF, 7% HCl, 37% H_2O
Ge	1	44% HNO_3, 26% HF, 27% CH_3COOH, 3% Br_2
	F2	90% HNO_3, 10% HF
Mg & alloys	1	2–15% HCl in H_2O or methanol
	F2	2–15% HNO_3 in H_2O or methanol
MgO		H_3PO_4, 100°C
Nb	F1	70% HNO_3, 30% HF
	2	34% H_3PO_4, 33% HF, 33% HNO_3
	3	12% HNO_3, 11% HF, 27% H_2SO_4, 50% H_2O
Si	1	33% HF, 17% HNO_3, 50% CH_3COOH
	F2	90% HNO_3, 10% HF
Silica glass	1	35% HF, 50% CH_3COOH, 15% HCl
	F2	5% HF, 2% HCl, 93% H_2O
Ta	1	50% HNO_3, 50% HF
	F2	56% H_2SO_4, 22% HNO_3, 22% HF

Table 3. *Cont.*

Material		Solution (use at room temperatures unless otherwise stated)
Ti–Nb	1	75% HNO_3, 25% HF
	F2	25% HNO_3, 25% H_2SO_4, 25% HF, 25% H_2O, 10°C
U & alloys		50% HCl, 50% H_2O
UO_2	1	25% HF, 25% HNO_3, 25% CH_3COOH, 25% chromic acid
	2	33% H_3PO_4, 33% HNO_3, 33% CH_3COOH
V		67% HF, 33% HNO_3
Zn & alloys	1	60% HNO_3, 40% H_2O
	F2	30–50% H_3PO_4 in ethanol
Zr, alloys	F	44% HNO_3, 12% HF, 44% H_2O

table can be found in the books of Goodhew (1972) and Thomas and Goringe (1980).

A few words of warning are necessary to those without experience in handling corrosive and oxidizing reagents, particularly those containing hydrofluoric and perchloric acids. Major points to be considered are listed in Table 4.

Table 4.

(1) Look up any unfamiliar reagent in the laboratory safety handbook or a source such as Alderson (1975) or the *Handbook of Laboratory Safety.*
(2) Remember that any reagent could be corrosive, poisonous, or explosive. Many electrolytes are unpleasant to inhale.
(3) Always add acid to solvent, never solvent to acid.
(4) Hydrofluoric acid (HF) penetrates both clothing and skin and is extremely dangerous. It attacks glass and must be used in polyethylene containers. It is essential to wear plastic gloves and apron.
(5) Mixtures containing perchloric acid ($HClO_4$) are liable to explode on contact with oxidizable material (paper, wood, etc.) or even spontaneously. Never keep these solutions for long, even in a refrigerator.
(6) Always label each solution with its composition and date.
(7) Dispose of any unlabelled solutions.
(8) Consult the laboratory supervisor to discover the procedures for disposal of waste solutions.
(9) Use fume cupboards whenever possible.
(10) In any case of doubt, ask for advice.

A particular problem with chemical thinning is stopping the reaction. Most solutions act quickly enough for the slice to be held with tweezers (preferably plastic!) so that it can be withdrawn from the solution to check that its faces are being polished rather than etched. A large beaker of solvent should be present beside the beaker of solution so that the slice can be immersed instantly to quench the chemical action. It is necessary to determine by trial and error how frequently the slice should be examined because in some cases it takes some time for the polishing action to be set up so that withdrawing the slice becomes counterproductive if carried out too often.

Table 5. *Names and formulae of reagents used in chemical and electrochemical thinning*

Name	Formula	Name	Formula
Acetic acid	CH_3COOH	Nitric acid	HNO_3
Acetic anhydride	$(CH_3CO)_2O$	Ortho-phosphoric acid	H_3PO_4
Cadmium hydroxide	$Cd(OH)_2$	Perchloric acid	$HClO_4$
Chromium trioxide	CrO_3	Potassium ferrocyanide	$K_4Fe(CN)_6$
Ethanol	C_2H_5OH	Sodium hypochlorite	$NaOCl$
Hydrochloric acid	HCl	Sodium hydroxide	$NaOH$
Hydrofluoric acid	HF	Sulphuric acid	H_2SO_4
Hydrogen peroxide	H_2O_2	Water	H_2O
Methanol	CH_3OH		

2.4. Making a disc

Many automated final thinning techniques require a disc specimen 3 mm in diameter. Such a disc is easily handled, fits directly into the microscope without a grid, and provides good structural support for the thinnest areas of the specimen. It is occasionally possible to prepare the material initially as a rod of diameter 3 mm from which discs can be cut using a small slitting wheel or a dedicated mini-cut-off machine (see Appendix for suppliers). Such discs will usually be about 1 mm thick and can be thinned further using any of the techniques described in the previous section before final thinning. It is more usual, however, to dish ('dimple') them in the middle so that the outer rim remains 1 mm thick and can be handled with tweezers while the centre is $100\,\mu$m thick and is ready for final thinning.

Dimpling can be done by electropolishing or by ion bombardment, but in both cases it needs to be done more rapidly than final thinning or it will take too long. It is not necessary for the dimple to have a perfectly polished surface, and thus the dimpling procedure does not have to be as carefully controlled as does the final thinning. Automated electropolishers, such as the Tenupol described in the next chapter, generally have an operating manual which suggests appropriate electrolytes for rapid dimpling. Alternatively a specially designed ion-bombardment source (such as the Ion Tech 'Microrapid') which provides a high ion beam current and thus a fast thinning rate can be purchased. The temperature rise of the specimen must be watched, however, since a great deal of energy is concentrated in the disc by the ion beam. Ion thinning devices are described in more detail in Chapter 3.

If the material has been prepared as a slice then the 3 mm disc must be cut from a sheet of thickness about $100\,\mu$m. The quickest way to do this, if the material is ductile, is to punch out the disc using a hollow punch of internal diameter 3 mm. Such a punch needs to have some device for holding the sheet flat but this is not difficult to arrange. A simple jig made in the author's laboratory is shown in Fig. 4, but similar devices are also available commercially (Appendix). Clearly the punching action should not appreciably deform more than a thin rim of material

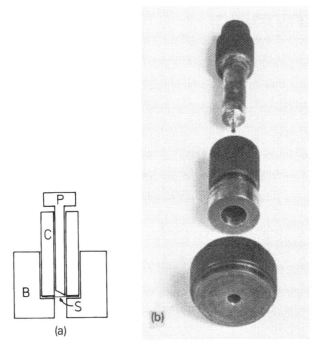

Fig. 4. A disc punch. The sheet specimen (S) is held down onto a base (B) by a block (C). The punch P runs through C and into a hole in the case when it is struck sharply with a hammer.

around the edge of the disc. This generally proves to be the case for ductile metals but the punching technique is not suitable for brittle materials. Unexpected damage may occur – it has been reported recently that after punching apparently successful discs from a steel, the thin area contained stress-induced martensite.

All the gentler techniques for cutting discs from sheet take much longer. The most commonly used methods use hollow drills. Diamond-tipped hollow drills can be purchased for use in conventional drilling machines or alternatively a tube of inside diameter 3 mm can be used in a spark machine. Finally, if the material is hard an ultrasonic drill with a hollow tool can be used. In such a machine the hollow tool is pressed very lightly onto the specimen sheet and is ultrasonically vibrated while immersed in a slurry containing abrasive powder. This works very well for silicon, for example, but is unsuitable for f.c.c. metals which simply have a circular groove forged into them.

The end product of initial preparation should be a parallel-sided sheet, a thin disc, or a dimpled thick disc. The techniques by which these can be finally thinned are the subject of the next chapter.

3

Final thinning

There are many ways of thinning a sheet or disc specimen down to its final electron-transparent thickness. Outside the biological sciences, however, two techniques dominate: these are electropolishing and ion beam thinning. This chapter therefore starts with a detailed description of methods based on these two approaches, continues with a shorter discussion of the less frequently used techniques such as chemical thinning and cleavage, and ends by considering how to thin specific regions of a specimen.

3.1. Electropolishing

The principles

The principle of electropolishing is quite simple. An electrolytic cell is established with the specimen as the anode and an appropriate potential is applied so that the specimen is dissolved in a controlled manner. Electropolishing is usually continued until a hole has formed in the specimen ('perforation'), when the regions around the hole should be thin enough for TEM. This sounds simple, and is simple once reproducible conditions have been established. However, there are so many experimental variables such as cell geometry, applied potential, and the composition, temperature, and velocity (if it is stirred or pumped) of the electrolyte, that some insight is needed into what should be happening in order to obtain good results quickly.

First consider what must be achieved. The electropolishing cell must polish the specimen, i.e. remove very fine-scale irregularities, it should smooth the specimen, i.e. remove larger-scale irregularities, and it must thin the specimen uniformly and fairly rapidly. The sequence is illustrated schematically in Figure 5. In order to produce this electropolishing action the electrolyte must generally contain an oxidizing agent together with reagents which will form a thin but stable viscous film. The fine polishing action is achieved by dissolution controlled by the length of the diffusion path through the viscous film to the electrolyte as shown in Fig. 6. High spots dissolve faster since they are nearer the free electrolyte and this results in a fine-scale smoothing which can generally be recognized as a brightening of the surface.

The viscous film must be kept thin and hence the electrolyte must contain, as well as an oxidizing agent and a film former, a solvent for the oxide-containing viscous film. Sometimes one reagent will act in all three ways and the electrolyte can be simple. An example is a dilute solution of perchloric acid in ethanol, which is close to being a universal electropolishing agent. On the other hand some

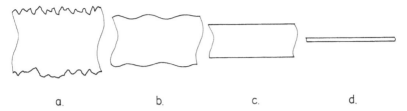

Fig. 5. The stages of electropolishing. The rough specimen (a) must be polished (b), smoothed (c), and uniformly thinned (d).

electrolytes (see Table 6 on p. 19) are quite complex mixtures of three or four reagents. In these cases an oxidizing agent such as perchloric or nitric acid, a film former such as phosphoric acid, another acid such as sulphuric acid to dissolve oxides, and a diluent (perhaps also viscous, like glycerol) to control the rate of reaction can be identified.

Once the composition of the electrolyte is determined the major variable is the applied potential. Generally too low a potential will lead to etching, while too high a potential will lead to pitting and thus uneven polishing. Both of these conditions are to be avoided and in principle the correct conditions can be determined from an experimental current—voltage curve. In a stable electrochemical cell a current—voltage curve should appear as in Fig. 7, curve a. The best polishing action then occurs in the plateau region. However, a potentiostat is necessary to measure a reliable current—voltage curve. Less sophisticated attempts to plot an experimental curve are usually dogged by difficulties associated with achieving stable steady-state conditions. A real experiment is more likely to lead to curve b in Fig. 7, which is not so helpful. The normal heuristic approach is therefore to start with the potential recommended in the recipe (see later this chapter) and then to increase this if etching occurs or decrease it if pitting occurs. This is not very scientific but it generally works!

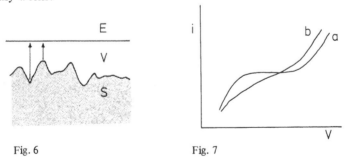

Fig. 6   Fig. 7

Fig. 6. Electropolishing: a viscous film (V) forms between the specimen (S) and the electrolyte (E); high spots have a shorter diffusion path through the layer and polish faster.
Fig. 7. Current—voltage curves for electropolishing. Ideally the curve a is followed, and the best polishing condition is on the central plateau. More commonly a curve such as b is found.

The practice

In order to start polishing three essential components are required: an electrolyte (chosen from Table 6), a cell (manual or automatic), and a power supply (usually d.c.). The power supply is the most easily dealt with (see Fig. 8). It needs to provide 5–50 V d.c., reasonably smoothed, at currents of up to 1–2 A, and this can be achieved using car batteries and a rheostat or by a variety of more expensive dedicated electronic power supplies. If there is nothing appropriate to hand suitable sources can be bought from any supplier of electropolishers (see Appendix).

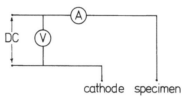

Fig. 8. The simplest power supply for electropolishing. A d.c. potential is applied across the electrolytic cell and its behaviour is monitored by a voltmeter and an ammeter.

The choice of electrolyte and cell depends on the specimen configuration planned. One way of thinning a fairly large sheet specimen (say 10 mm × 10 mm) and one technique for jet thinning a 3 mm disc specimen are described below.

3.2. Thinning a sheet: the window technique

The most straightforward (and cheapest) polishing cell is simply a beaker of electrolyte into which the cathode and the specimen are dipped. The cathode is a sheet of either the same material as the specimen or an inert material such as platinum. The specimen acts as the anode and a sheet 10–20 mm square is generally held in metal tweezers so that the potential can be applied via crocodile clips. Figure 9 shows the layout. The edges of the specimen, and the tweezers, must be protected from attack and this is achieved by painting them with an acid-resistant lacquer such as Lacomit. A 'window' of metal remains exposed, which gives the technique its name (Fig. 9, inset). The tweezers are held by hand so that the specimen is submerged, and the potential is applied. The advantages of this approach are that the specimen can be viewed while polishing is taking place since most electrolytes are transparent, and the polishing action can be stopped very quickly by switching off the power supply and immersing the specimen in an adjacent beaker of solvent. The power supply should always be switched off before removing the specimen from the electrolyte since otherwise it is possible for a spark to pass to the electrolyte as the specimen is withdrawn. In the majority of cases this is harmless, but of course it may cause ignition of flammable or explosive electrolytes and it is therefore good practice never to allow it to happen.

All the electropolishing variables can be controlled quite easily using the window

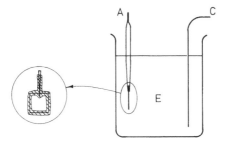

Fig. 9. The window technique: the sheet specimen is held in tweezers and lacquered; it then forms the anode in the electrolytic cell.

technique. The electrolyte can be cooled by standing the beaker in a trough of coolant or by pouring liquid nitrogen over the top of the electrolyte. The solution can be stirred using a magnetic stirrer and additional agitation can be introduced by moving the specimen to and fro.

The most crucial part of the window technique is the handling of the specimen. The lacquer must be applied carefully, only round the edges, and must be left to harden for several hours, although this can be accelerated by warming the specimen in hot air. The procedure is then as illustrated schematically in Fig. 10. The specimen is polished until perforation occurs, probably at the top of the window. If a loose film forms during polishing it can often be removed by raising and lowering the specimen through the electrolyte surface until it floats away. After perforation of a polished specimen the exposed edge should be examined by eye or hand magnifier. If it is smooth (Fig. 10(b)) then it should be covered with lacquer and polishing restarted (Fig. 10(c)). Some authorities recommend turning the specimen upside down at this stage but this is tedious since the tweezers have to be cut off and re-attached with lacquer. It is often possible to omit this step. When perforation occurs again, the edge should look more ragged (Fig. 10(d)). If so, the specimen is probably ready (Fig. 10(e)). If the edge is not yet ragged, relacquer and repolish until it is.

The specimen must now be washed thoroughly. This is a very important step and

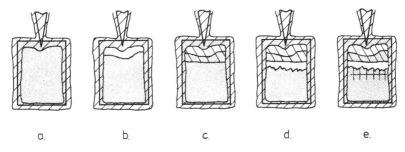

a.   b.   c.   d.   e.

Fig. 10. Electropolishing using the window technique: (a) the window is painted on; the specimen is (b) polished to perforation, (c) relacquered and (d) repolished; (e) foils about 2 mm square are cut from the jagged edge.

should be repeated several times, starting with the base solvent for the electrolyte being used and ending with clean alcohol. It is possible to remove the lacquer using acetone, but it is difficult to remove all traces of it and there is a tendency for it to redeposit over the whole specimen. For this reason the lacquered regions are best left in place or cut away using an old scalpel blade or miniature scissors. Several thin foils can now be cut from the specimen using rolling strokes with a new curved scalpel blade along the grid of lines in Fig. 10(e). This should be done with the specimen under alcohol. The foils (2 mm x 2 mm) can now be handled using a fine paint brush. As the alcohol dries the foil will be easy to detach from the brush and position on a support grid (see Chapter 5).

The advantages of the window technique are that it is controllable, simple, cheap, and produces several specimens at once. On the other hand the specimens need support in the microscope, are easily damaged, and the process is not highly reproducible. Some measure of 'craft skill' seems to be needed to produce foils repeatedly using the window technique. For this reason many laboratories find it worthwhile to invest in an automatic polisher such as that described in the next section. Such systems tend to be much more reproducible and less 'user specific'.

3.3. Thinning a disc: the automatic jet polisher

Many automated electropolishers have been devised and made in the last 20 years. Most of them have worked on the principle that a jet of electrolyte is directed at the centre of a 3 mm disc specimen in order to accelerate the attack and thus ensure perforation at the centre before the edge is appreciably thinned. This produces an ideal specimen for mounting in the microscope, with robust edges which can be handled and which serve to support the thin areas in the centre. One particular commercial device, the Tenupol (Struers Ltd.), has proved to be more popular than any other in the last few years and we shall therefore describe its use in some detail. However, most of what follows is applicable to other automated jet polishers – only the exact geometry differs.

The Tenupol is a double-jet device which thins both sides of the specimen at once. Figure 11 is a very schematic diagram of the polishing cell. A pump (in the electrolyte reservoir below the cell) circulates electrolyte through the dual jets which are directed at the centre of the 3 mm disc specimen. This is mounted in a removable holder (Fig. 12) and the anode potential is applied via a metal strip which runs through the centre of the holder. The polishing action needs to be stopped as soon as a hole has formed in the disc specimen. If it is carried on beyond this point then the thinnest areas can be rounded off by further electrolytic attack and/or the mechanical action of the liquid jets can deform the thin regions. Neither effect is desirable so the moment of perforation is usually detected using a light source on one side of the cell and a sensitive photodiode on the other (see Fig. 11). In favourable conditions perforation can be achieved in only a minute or two and the time taken is extremely reproducible, which is a great advantage for the microscopist.

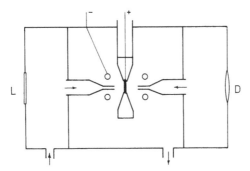

Fig. 11. A schematic diagram of the Tenupol (Struers Ltd). The electrolyte is pumped up from the tank below the cell and directed at the disc specimen by two jets. Perforation can be detected by a photodiode at D which registers light from the lamp L as soon as a hole forms.

There are a number of experimental variables associated even with an automated polisher. The electrolyte, its temperature, the applied potential, and the pumping speed must be chosen. Struers supply their own electrolytes with code numbers such as D-2 and recommend in their instructions the appropriate electrolyte for thinning a number of metals. However, it is often cheaper, and certainly more appropriate for materials not listed in their literature, to select an electrolyte from Table 6. It is wise not to use electrolytes containing HF, however, because of both its corrosive action on some components of the cell and the difficulty of avoiding minor leakage while using the Tenupol. The electrolyte can be cooled either by passing a coolant through the pipes provided in the Tenupol or more crudely by pouring liquid nitrogen onto the electrolyte in the reservoir before placing the pump and cell on top. The electrolyte volume is large, the reservoir is well insulated, and because polishing is generally quite quick a single application of liquid nitrogen will last for several specimens and continuous cooling is generally not necessary. As a rule, the electrolyte should be cooled if the recipe recommends it or if polishing is too fast to control at room temperature.

The electrolyte flow rate is an important, but difficult, parameter. Too fast a flow breaks up the viscous film and destroys the polishing action. It may also deform the foil after perforation. However, too slow a rate will not give good localized attack at the centre of the disc. The right conditions must be found by trial and error but should be reproducible for each combination of material and electrolyte.

The same care should be taken in washing the thinned specimen as was emphasized in the previous section. Fortunately a disc specimen is easily handled with tweezers by its rim and should need neither further trimming nor supporting in the microscope.

3.4. Recipes for electropolishing

Several electrolytes are given for each material (wherever possible) since ideas on how to vary the composition slightly for a slightly different alloy or heat

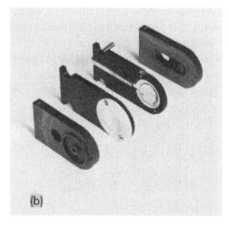

Fig. 12. The Tenupol automatic jet electropolisher showing the specimen holder (a) in place and (b) removed and dismantled.

treatment may be needed. Details of the original literature sources for many of these electrolytes can be found in the book by Goodhew (1972). If no potential is quoted, try about 10 V at first and then adjust. Many of the perchloric acid–alcohol mixtures which were originally used at concentrations of about 20 per cent $HClO_4$, 80 per cent C_2H_5OH can be used successfully at much lower concentrations such as 5 per cent or 10 per cent which result in much safer solutions. Solutions prefixed with a T are known to work in the Tenupol and many others are likely to work.

The electrolytes listed in Table 6 may work best when freshly made up or in some cases may improve when kept for a week or two. Some even improve after

Table 6

Material		Electrolyte
Al (aluminium) and many alloys		Most electrolytes are based on perchloric acid–ethanol mixtures; beware explosion hazard
	T1	10–20% $HClO_4$, 70–80% ethanol, $-20°C$
	2	8% $HClO_4$, 11% 2-butoxyethanol, 79% ethanol, 2% H_2O, 15°C
	3	33% $HClO_4$, 67% $(CH_3CO)_2O$, 20 V, hazardous
	4	7% $HClO_4$, 11% ethylene glycol, 9% H_2O, 73% ethanol
	5	62% H_3PO_4, 14% H_2SO_4, 24% H_2O + 160 g l^{-1} CrO_3, 10 V, 70°C
	6	40% CH_3COOH, 30% H_3PO_4, 20% HNO_3, 10% H_2O, $-10°C$
Al-Zn	T7	3% $HClO_4$, 32% butoxyethanol, 65% methanol, 25 V, $-30°C$, slow flow
Be (beryllium)	1	18% $HClO_4$, 18% 2-butoxyethanol, 64% H_2O, $-60°C$, 10 V
	2	60% H_3PO_4, 35% glycerol, 5% H_2O + 1 g l^{-1} CrO_3, 45°C, 35 V
Bi (bismuth)	1	Saturated solution of NaCl in H_2O + 2% HCl, $-10°C$, 5 V
Bi_2Te_3	2	5% NaOH, 4% tartaric acid, 53% H_2O, 38% glycerol, 10 V
Cd (cadmium)	1	44% H_3PO_4, 56% distilled H_2O, Ni cathode
	2	10% HNO_3, 90% CH_3OH
	3	20% $HClO_4$, 10% glycerol, 70% CH_3OH, 20 V
Co (cobalt) and alloys	1	23% $HClO_4$, 77% CH_3COOH, 22 V
	2	50% H_3PO_4, 50% H_2O
	3	14% HCl, 86% CH_3OH, 40 V
	4	2% $HClO_4$, 8% citric acid, 10% propanol, 80% C_2H_5OH + 50 g l^{-1} sodium thiocyanate.
	5	22% $HClO_4$, 14% butoxyethanol, 64% C_2H_5OH, 10 V
	6	20% HCl, 80% H_2O saturated with NaCl
Cr (chromium) and alloys	1	10% $HClO_4$, 90% CH_3OH, $-30°C$, 10 V
	2	44% CH_3COOH, 44% 2-butoxyethanol, 12% $HClO_4$, 35 V
	3	20% $HClO_4$, 80% CH_3COOH, 8°C, 12 V
Cu (copper) and alloys	1	33% HNO_3, 67% CH_3OH, 5 V
	2	66% H_3PO_4, 34% H_2O; wash in 10% H_3PO_4, then in H_2O, and then in CH_3OH
	3	40% H_3PO_4, 30% CH_3OH, 30% H_2O
	4	25% $HClO_4$, 75% CH_3COOH
Cu_3Au	5	95% CH_3COOH, 5% H_2O + 150 g l^{-1} CrO_3, 25 V
Fe (iron) and steels	T1	5% $HClO_4$, 95% CH_3OH, $-50°C$, 60 V
	2	5–10% $HClO_4$, 95–90% CH_3COOH
	3	95% CH_3COOH, 5% H_2O + 400 g l^{-1} CrO_3
	4	54% H_3PO_4, 36% H_2SO_4, 10% H_2O
	5	H_3PO_4 + excess CrO_3
Ge (germanium)		0.05 M KOH
Hg (mercury)		20% HNO_3, 80% CH_3OH, $-60°C$, keep cool
In (indium)		33% HNO_3, 67% CH_3OH
Ir (iridium)		Saturated $CaCl_2$ in H_2O + 1% HCl
Mg (magnesium)		Mg itself is normally chemically polished in $HNO_3 - CH_3OH$ MgO is chemically polished in H_3PO_4
	1	33% HNO_3
	2	20% $HClO_4$, 80% C_2H_5OH, $-10°C$, 10 V
	3	1% $HClO_4$, 99% C_2H_5OH, $-55°C$
Mo (molybdenum)	1	14% H_2SO_4, 86% CH_3OH, 10 V
	2	15% HNO_3, 85% H_2O, 35 V

Table 6. *Cont.*

Material		Electrolyte
Nb (niobium)	1	10% HF, 90% H_2SO_4, 60°C
	2	10% HF, 13% HNO_3, 77% H_2SO_4, 65°C
	T3	0.05 mol l^{-1} $Mg(ClO_4)_2$ in CH_3OH, -5°C, 60 V
	4	10% $HClO_4$, 90% CH_3OH, -30°C
Ni (nickel)	1	20% $HClO_4$, 80% C_2H_5OH, 0°C, 7 V
	2	23% $HClO_4$, 77% CH_3COOH, 25 V
	3	10% $HClO_4$, 90% CH_3COOH + 20 g l^{-1} CrO_3 + 10 g l^{-1} $NiCl_2$, 60 V
	4	10% H_2SO_4, 10% H_3PO_4, 60% glycerol, 20% H_2O, 15 V
	5	57% H_2SO_4, 43% H_2O, 10 V
	6	6% H_2SO_4, 94% CH_3OH, -70°C
Pd (palladium)	1	33% HNO_3, 17% H_3PO_4, 50% CH_3OH
	2	33% H_2SO_4, 33% H_3PO_4, 34% HNO_3
	T3	0.5 mol l^{-1} LiCl + 0.2 mol l^{-1} $Mg(ClO_4)_2$ in CH_3OH, room temperature
Pt (platinum)	1	Saturated solution of $CaCl_2$ in H_2O + 2% HCl
	2	33% H_2SO_4, 33% H_3PO_4, 34% HNO_3
Rh (rhenium)	1	20% $HClO_4$, 80% C_2H_5OH, -40°C, 20 V
	2	3% NaOH, 97% H_2O, room temperature, 200 V
Sn (tin)	1	5% $HClO_4$, 10% 2-butoxyethanol, 85% C_2H_5OH
Ta (tantalum)	1	5% H_2SO_4, 1.25% HF, 93.75% CH_3OH, 0°C, 60 V
	T	90% H_2SO_4, 10% HF
Ti (titanium) and alloys	1	5% H_2SO_4, 95% CH_3OH, wash in HNO_3
	2	5% $HClO_4$, 95% CH_3COOH, 40 V
	3	5% $HClO_4$, 35% 2-butoxyethanol, 60% CH_3OH, -30°C, 30 V
U (uranium)		There is often a problem in removing the oxide film and preventing staining
	1	33% H_3PO_4, 33% H_2SO_4, 34% H_2O, 75 V
	2	20% $HClO_4$, 80% C_2H_5OH, -50°C, 10 V
	3	80% CH_3COOH, 20% H_2O + 150 g l^{-1} CrO_3, 20 V
V (vanadium) and alloys	1	15% H_2SO_4, 85% CH_3OH, wash in the same solution
	T2	20% NaOH in H_2O
W (tungsten) and alloys WC + Co	1	1.5% NaOH, 98.5% H_2O, 20 V
	2	5% H_2SO_4, 1.25% HF, 93.75% CH_3OH, 0°C, 60 V
	3	20 g l^{-1} NH_4Cl, 75 g l^{-1} tartaric acid, 20 g l^{-1} NaOH, 40 g l^{-1} Na_2CO_3, 5 g l^{-1} $NaNO_2$
Zn (zinc)	1	20–30% HNO_3, 80–70% CH_3OH, 5 V
Zr (zirconium) and alloys	1	10% $HClO_4$, 90% C_2H_5OH, -30°C, 20 V
	2	10% $HClO_4$, 90% CH_3COOH
	3	10% HF, 45% HNO_3, 45% H_2O

having been used. It is therefore worth trying both old and new electrolyte. However, this does place a premium on systematic laboratory technique. A detailed notebook in which experimental work is recorded should of course always be kept, but in addition it is important to label all bottles of electrolyte with the composition, the date and the experimenter's name. Unlabelled or very old solutions should be disposed of, and will be by most laboratory supervisors. Note that gold, silver and their alloys are omitted from Table 4; these usually require electrolytes containing potassium cyanide, which is very poisonous, and should never be used except under the most strictly controlled conditions and with the knowledge and advice of the laboratory safety officer.

There are a few general points about electropolishing which deserve mentioning before we leave the subject. The need for thorough washing of the specimens cannot be emphasized too strongly. This will generally involve at least four or five clean dishes of solvent and should end up with some rapid-drying non-staining solvent such as alcohol.

Many electrolytes work better when cooled. There are a variety of ways of achieving this, the simplest being to stand the electropolishing cell in a trough of cooled liquid. This could be water plus ice, alcohol plus dry ice (solid CO_2), or alcohol with liquid nitrogen poured on top. It is possible to buy small refrigerators which will pump cooled liquid through a coil immersed in the electrolyte (as in the Tenupol) but this is an expensive expedient compared with pouring liquid nitrogen on top of the electrolyte.

3.5. Ion beam thinning

Theory

The principle of ion beam thinning is extremely simple. A beam of inert gas ions or atoms is directed at the (usually disc) specimen, from which it removes surface atoms in a process known as sputtering. If this can be achieved without the creation of artefactual damage then ion beam thinning is an ideal method for the preparation of foils from both conducting and non-conducting materials. However, it is necessary to anticipate and control several potentially undesirable effects. These include the implantation of the sputtering ion, the development of rough surface topography, and the heating of the specimen. For these reasons it is necessary to control the nature of the ions, their energy and direction of incidence, and their frequency of arrival (i.e. the ion beam current). We shall first consider the effect of each of these variables on the sputtering process and then look at some widely available pieces of thinning equipment.

Sputtering will occur when any ion carrying more than about 100 eV of energy hits a surface. The number of atoms ejected by each incident ion or atom is known as the sputtering yield Y. In general Y and hence the thinning rate increases as the ion energy increases and as the mass of the bombarding ion increases, but Y decreases as the atomic mass of the specimen increases. Our desire for a high sputtering yield but no chemical change in the specimen dictates the use of argon: lighter inert gases (helium and neon) thin more slowly and heavier inert gases (krypton and xenon) are too expensive. The ion energy is also quite easy to select. At first Y increases with ion energy (Fig. 13) but eventually at high energies the incident ion is deposited far below the surface and fewer surface atoms are ejected. The optimum energy is thus somewhere in the range 1–10 keV, and a value between 3 and 6 keV is generally used. The sputtering yield also depends on the angle at

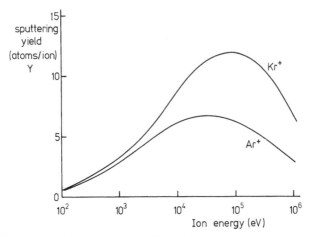

Fig. 13. Ion beam thinning: the sputtering yield rises with the energy of the ions until a peak is reached (heavier ions sputter more efficiently).

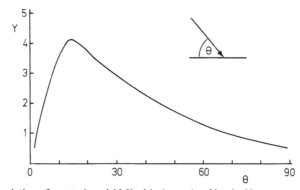

Fig. 14. The variation of sputtering yield Y with the angle of ion incidence.

which the ions hit the surface (Fig. 14). For this reason (and others (see next paragraph)) a glancing angle of incidence between 5° and 30° is generally used.

One of the most difficult problems associated with ion beam thinning is the development of a rough surface. As we have seen earlier (Chapter 1) the final foil should be parallel-sided. However, if a fixed angle of glancing ion incidence is maintained, surface corrugations develop on most materials. The most common method of minimizing this effect is to rotate the specimen while thinning.

Practice

Having decided on the ideal operating parameters we have to choose the appropriate hardware. There are two main types of ion gun in use, each of which suffers from inconveniently rapid wear because the ions erode the gun as well as the specimen. Regular replacement of gun components (particularly the cathode) is thus an inevitable feature of ion beam thinning. The two ion sources are illustrated schematically in Fig. 15. In the historically earlier hollow anode gun (Fig. 15(a)) the gas is ionized between the anode (A) and cathode (C) or within the anode, and is then accelerated through the hole in the cathode. This gun is very simple but delivers quite a small current into a rather divergent ion beam. This leads to slow thinning but minimizes heating of the specimens. The saddle field gun (Fig. 15(b)) is more complex but emits two very fine and very intense beams of ions (or a mixture of ions and neutral atoms). This can lead to much faster thinning but also transfers more energy to the specimen and may therefore heat the specimen too much if the ion beam current is not carefully limited.

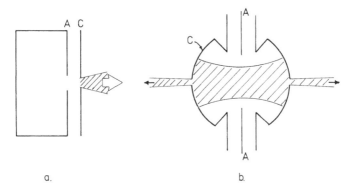

Fig. 15. (a) The hollow anode ion gun, which emits a single divergent beam of ions; (b) the saddle field gun which emits two fine beams of ions and atoms. In each diagram A and C indicate the anode and cathode.

Commercial thinning machines using each type of ion gun are available (see Appendix). In addition to facilities for setting the ion energy and beam current of at least two guns these machines generally incorporate a specimen holder which accepts 3 mm discs and rotates them about their perpendicular axis while they are thinned from both sides simultaneously. Figure 16 shows a typical layout and also indicates how perforation of the specimen can be detected using either light or a third ion beam. It is particularly useful to have automatic termination of thinning since erosion is generally slow (between one and a few tens of microns per hour) and therefore the thinning of a single $50\,\mu$m thick disc can take several hours. For the same reason it pays to prepare the disc as thinly as possible (consistent with not damaging it and being able to handle it) before starting ion beam thinning. The large ion current which it is possible to draw from a saddle field source has led to the development by IonTech Ltd. of a rapid ion pre-thinning device (Microrapid) which permits erosion at more than $100\,\mu$m h^{-1} for most materials by placing the specimen very close to the ion source. This is fine for refractory materials but great care must be taken that the specimen temperature does not rise to an unacceptable level during such pre-thinning.

Ion beam thinning has become widely used and is most useful for ceramics,

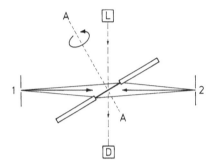

Fig. 16. The specimen in an ion beam thinner is usually held at a glancing angle to the beams from two ion guns 1 and 2. The specimen is rotated about the axis A−A. Perforation can be detected by a light or third ion gun at L and a suitable detector at D.

glasses, and other non-conductors and for two-phase materials in which one phase always tends to electropolish faster than the other. Not all materials sputter at the same rate under a given set of ion beam conditions, but when the ion beam is incident at a glancing angle to a rotating specimen differential thinning is minimized.

The technique is least useful for light ductile conducting crystals where the microstructural damage which is introduced and the gas atoms which are implanted close to the surface may lead to confusing artefacts in the micrographs.

3.6. Other final thinning techniques

Electropolishing and ion beam thinning probably account for more than 90 per cent of all final thinning for non-biological microscopy. However, there are a number of other techniques in use for particular materials and we shall mention some of these briefly in this section. We have already seen (Chapter 2) that chemical dissolution is widely used to pre-thin sheet material before it is finally thinned. In a few cases chemical thinning can successfully be continued until the sheet perforates in order to produce thin foils directly. As with electropolishing, this can be done in a variety of ways, of which the simplest merely involves holding the sheet (about 15 mm square) with tweezers in a beaker of the polishing solution − the sheet may be lacquered round the edges, as in the window technique (p. 14) or may be unprotected, in which case it is generally a good idea to immerse it only partially in the solution so that attack occurs preferentially at the 'water-line'. These simple techniques are quite successful for magnesium and MgO for example.

Silicon is generally prepared by chemical thinning, and in this case it is very desirable to have a disc specimen since thin silicon is very fragile. Chemical thinning of a 3 mm diameter disc cannot easily be done by holding it with tweezers so an arrangement such as that shown in Fig. 17 is used. The thinning solution is dripped onto the centre of the disc, which must be rotated in order to avoid the development of a 'run-off' channel. A minor problem with this set-up is that it must

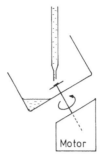

Fig. 17. The arrangement for chemically thinning silicon, for example. Acid is dripped from a pipette onto the centre of a rotating disc specimen. A beaker catches the spent acid.

generally be constructed of polyethylene or polytetrafluoroethylene (PTFE) since the best polishing solutions contain HF.

Polishing solutions which have been found to be successful for final thinning are those marked with an F in Table 3.

A few materials exist as a layer structure and often cleave along one particular crystallographic plane. This fact can in many cases be exploited to produce thin foils. A freshly cleaved surface will often yield thin foils as loose fragments which can either be picked up directly or can be dislodged by pressing and peeling off a piece of adhesive tape. This is of course a very quick way of generating TEM specimens but it only works for a limited range of materials, e.g. graphite, and it only permits a single orientation of specimen to be observed.

The final two techniques to be considered can both be called 'mechanical thinning' and consequently carry the danger that specimens will be damaged by the thinning process. However, very hard materials (carbides, some oxides, etc.) can be mechanically polished, using fine diamond or SiC powders, to electron transparency. This can be achieved by using a polishing jig such as that shown in Fig. 3, or by a variety of embedding techniques. The most difficult practical problem, however, is to remove the thin specimen from the adhesive or embedding medium which is necessary to secure it to a mount so that you can hold it. Most adhesives swell under the action of solvent or heat and this can be sufficient to break the specimen before it is released. Mechanical polishing is thus only used when all else fails.

The second 'mechanical' technique is the ultramicrotome, which is a fine slicing machine widely used by biologists to cut tissue sections embedded in epoxy resin. The same technique of embedding in resin followed by microtoming can be used for a wide range of 'difficult' non-biological materials. We have successfully employed ultramicrotomy to prepare thin foils from porous materials such as cement and bone, from particles including catalysts, and from polymeric materials including textile fibres. The techniques of embedding and cutting are skilled and complex, and consultation with a handbook on biological specimen preparation and a practising microtomist (courtesy of the nearest biological electron microscopy unit) is recommended.

3.7. Thinning specific regions of a sample

There are many occasions when it is necessary to examine a specific part of a larger specimen. When this is a grain boundary, a phase boundary or a precipitate particle there is little alternative to thinning several specimens and hoping that the desired feature will appear in the thin region. With larger microstructural features it can be ensured that the selected area is in the centre of the 3 mm disc before final thinning. However, this is not adequate for investigation of a region which does not normally survive thinning such as that just beneath one surface of the sample or the material at the root of a crack. Let us consider two of the more common types of procedure.

3.8. Thin foils from just beneath the surface

The essential requirement here is that final thinning be from one side only. It is quite easy in principle to cut a disc or sheet such that one face is the original sample surface and then electropolish from the other side until perforation. The thin region should then be as near as possible to the original surface, which might be a fracture surface for instance. However, in practice this is not so easy. It is possible to lacquer the side to be preserved and electropolish the other using either the window technique or a jet technique, but it is very difficult to prevent the erosion of the protected surface immediately after perforation and before the specimen is out of the electrolyte and washed. Similarly in a twin-jet electropolisher it is difficult to blank off one jet *perfectly* for single-sided thinning. The best approach is to let any protective lacquer dry really hard before starting single-sided thinning and to watch carefully for perforation and then wash the specimen as soon as possible.

If single-sided thinning is to be a recurrent requirement it might be worth investing the time to construct a special electropolishing cell of the type shown in Fig. 18. We have used this very successfully over a number of years. The key feature is that the protected surface of the disc specimen is stopped off with a dense inert liquid (e.g. dibromoethane). As soon as perforation occurs this seeps through the hole and automatically stops both the electropolishing action and any residual chemical attack. This action can be controlled by raising or lowering the liquid reservoir and thus adjusting the pressure forcing the stop-off liquid through the hole.

This type of single-sided thinning can also be used to reveal structure at a known depth beneath one surface of a specimen, for example when ion implantation or diffusion are being studied. The extra requirement here is for a layer of controlled thickness to be removed from the surface before thinning from the other side is started (Fig. 19). This layer can be removed by mechanical polishing for hard materials (for example a vibratory polisher can be calibrated for erosion rate) or by carefully controlled electropolishing or chemical polishing. It is worth

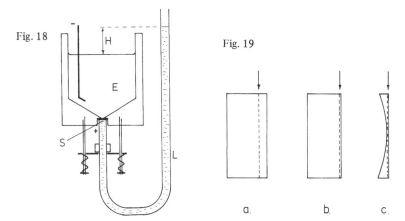

Fig. 18. A single-sided electropolisher. The specimen (S) is held by a spring against the electrolytic cell (E). Behind the specimen is an inert liquid (L) which will flow through the hole as soon as perforation is achieved. The pressure of liquid is controlled by adjusting the head (H). Stirring is often necessary and is introduced by a motor-driven paddle from above.

Fig. 19. If a specific subsurface region is to be examined (broken line in (a)) then a controlled thickness must be removed from the surface (b) and final thinning must be carried out from the back (c).

considering, for those materials where it is applicable, the use of anodizing since this gives a reproducible thickness of oxide each time the potential is applied. Layers from 10 nm to 1 μm can be removed in this way from aluminium, titanium, niobium, etc.

It is obviously quite easy to thin from one side using ion beam thinning simply by only operating one of the ion guns, but there is a hidden danger here. The sputtered material must redeposit somewhere and there is plenty of evidence that it will readily do so on the back of the specimen so that the original surface cannot be located precisely. A solution to this could be to lacquer the back of the specimen before thinning, as if it were to be electropolished. The redeposited layer should then dissolve off with the lacquer after thinning. Alternatively it may be possible to insert a blank piece of material on one side of the specimen.

3.9. Cross-sections perpendicular to the original surface

Very often it is required to know how the microstructure of the sample varies with depth below its surface. A cross-sectional specimen must be prepared in this case, but this is difficult because the area of greatest interest is then right at the edge of the specimen (Fig. 20(a)). The most satisfactory solution is to deposit additional similar material onto the surface of the sample, as shown in Fig. 20(b), before cutting the cross-section. The region of interest is now in the centre of the sheet and there is a reasonable chance both of thinning a disc specimen to reveal the original surface and of recognizing the original surface since it is unlikely that the structure of a deposited layer is the same as that of the specimen. The extra

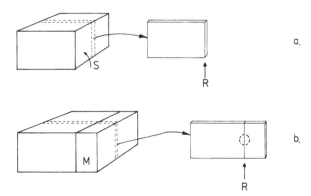

Fig. 20. If the structure of interest lies just beneath the surface S a conventional section (a) is no use since the region to be examined (R) is at the edge of the section. The solution (b) is to build up the specimen with deposited material (M) until the same section contains the region of interest in the middle. A disc can now be cut containing this region (R).

material can be deposited using electroplating, for most metals, or by vacuum deposition for a variety of materials. These techniques can also be used to fill up cracks when it is required to thin the material at the root. If the crack is not filled it is difficult to prevent preferential erosion at the tip of the crack which will mean that the thinned area will be some way from the original crack tip.

In addition to a few general references the Bibliography includes one example reference to each technique mentioned. These are intended as starting points, and not as a comprehensive survey of specialized techniques.

Finally a review of the most appropriate starting points for the common classes of materials is presented in Table 7.

Table 7. *Survey of final thinning techniques*

Type of material	Most likely technique
Metal (single-phase)	Electropolishing
Metal (two-phase)	Electropolishing; ion beam thinning
Semiconductor	Chemical polishing
Ceramic	Ion beam thinning; chemical polishing
Polymer	Ultramicrotomy

Replicas

4.1. The role of the replica in modern microscopy

The role of the replica has changed dramatically since the technique was established in the 1940s. Until the 1970s a replica was primarily a method of reproducing surface topographic detail so that it could be viewed in a transmission electron microscope. Many precise and sophisticated single-stage and two-stage techniques were developed for this purpose, the majority of which led to a thin carbon film in which topographic detail was made visible by 'shadowing' with a heavy metal (Fig. 21). Since we now have scanning electron microscopes available with a resolution better than 5 nm the need for this sort of replication has considerably diminished. Direct replication only survives in the materials sciences for a few special problems, such as the examination of the surface of a large component without cutting it up or the study of radioactive material which cannot itself be put in an ordinary unshielded microscope.

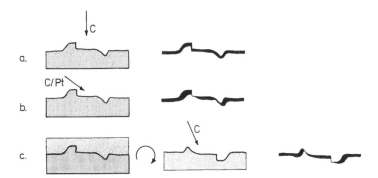

Fig. 21. Three replication techniques: (a) direct carbon replica; (b) shadowed carbon—platinum replica; (c) two-stage plastic—carbon replica.

A second type of replica using 'extraction' techniques is still used, however, and is indeed experiencing a revival of interest with the spread of 'analytical' transmission electron microscopes. Many problems involving the determination of the composition, crystal structure, or orientation of small second-phase particles are made much easier if the particles are extracted from their matrix and then supported in the microscope by means of a replica. Extraction replicas can preserve the relative positions and orentiations of second-phase particles as long as they are small enough to be supported by a carbon film (say smaller than 1 μm). It is therefore worth considering these replicas in detail.

4.2. The extraction replica

The general principles of the technique are shown in Fig. 22. The specimen is etched to make the particles of interest stand proud of the surface. A carbon coating is applied and the replica is then stripped off (perhaps using a second etch) carrying with it many of the second-phase particles. If a suitable etch has been found the number, shape, and distribution of particles will be preserved in the replica. We shall now consider the necessary practical steps in detail.

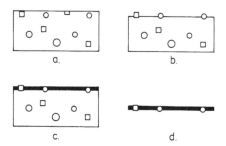

Fig. 22. The extraction replica. The two-phase specimen (a) is etched (b) and carbon coated (c). A further etch then lifts off the replica containing some second-phase particles (d).

A metallurgical specimen is generally first polished flat, in order to make the later lifting of the replica as easy as possible, and then etched. The etch may be chemical or electrolytic but should be chosen to remove the matrix while not attacking any particles of interest and should generally be quite shallow so that particles are exposed but not removed. The choice of etchant is determined by metallurgical considerations and a standard text such as Smithells (1976) should be consulted. However, dilute nital (2–10 per cent nitric acid in methanol) or dilute solutions of bromine in ethanol are effective for many alloys. The specimens must now be washed and dried and the replication stage should be started as soon as possible to avoid the deposition of airborne dust which would contaminate the replica.

A thin layer of carbon must now be coated onto the etched surface. This is generally achieved by using a carbon arc source in a vacuum coating unit. Many commercial designs are available (see **Appendix**) but all work on the principle shown in Fig. 23. Two carbon rods about 5 mm in diameter are lightly pressed into contact by a spring. The end of one rod is filed flat while the other is sharpened to a point using a device like a pencil sharpener. An arc is struck between the rods and (surprisingly to the uninitiated) a supply of virtually monatomic carbon is generated. This is deposited on every surface inside the vacuum bell jar including the specimen, which is usually sited so that the angle of incidence will be 45–60°. The critical parameter, once the arc has been struck, is the time of deposition

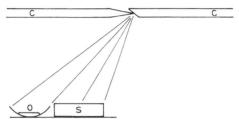

Fig. 23. Carbon coating: an arc is struck between two carbon rods (C); a fine carbon layer is deposited on the specimen (S) and on an oil drop in a porcelain dish (O).

since this will affect the thickness of the replica. Although more sophisticated techniques are available, such as quartz crystal thickness monitors, it is generally just as satisfactory to put beside the specimen a small piece of white porcelain with a drop of oil on it. As carbon is deposited the porcelain will discolour but the oil will not. It is normally satisfactory to stop coating when a distinct difference can be seen between the porcelain and the oil. However, this is a matter for trial and error and after a few attempts the oil may not be necessary. As a guide, in most circumstances the carbon arc will only be needed for a few seconds.

The next stage is to remove the specimen from the vacuum system and strip off the replica. Unless the specimen is very small or the replica is found to lift extremely easily, it will be best to score the carbon film with a sharp point into 2–3 mm squares. The specimen can then be lightly etched again, using the same etchant as before, until the replica begins to lift. It is important not to etch too vigorously since the formation of bubbles will probably break up the replica. Before the replica squares float off, remove the specimen from the etchant and slide it into a dish of water (Fig. 24). The replicas should now float off onto the water surface as the specimen is immersed. Leave them floating for a while to wash the electrolyte away, and pick them up on a 3 mm grid. This is done by gripping the edge of the grid with tweezers, plunging it under the water, and bringing the grid up underneath the floating replica square. With a little practice the replica can be picked up so that it remains centrally on the grid. The drop of water which will still be attached to the grid and the tweezers is best sucked up by sliding a piece of filter paper between the prongs of the tweezers. When it is dry the grid and its replica can be stored (Chapter 5) or put straight into the microscope. It is generally not a good idea to put a wet grid and replica down on a piece of filter paper. A thin replica will be damaged by the fibres of the filter paper.

For some diffraction studies it is helpful to have a crystallographic standard in the same plane as the specimen, and this can be achieved rather easily by evaporating a layer of gold or aluminium onto the replica. This can usually be done from a wire or a boat in the same coating unit as was used for making the replica. It may be useful to coat only half the replica in this way by shielding the other half of the grid with a piece of paper.

The extraction replica technique described here is not only applicable to metallurgical systems but can be used with slight modifications to prepare fine powders

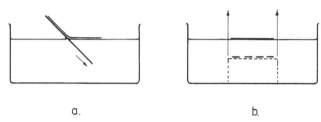

a.   b.

Fig. 24. (a) Lifting off a replica or carbon film. The substrate is slid gently into a water bath. The carbon will remain on the surface. If this does not work first time it can be repeated and the leading edge of the film can be teased with a needle. (b) Picking up a carbon film on grids. The 3 mm grids are arranged on a mesh beneath the floating film. The mesh is lifted out of the water bath through the film, which will adhere to the grids. Each grid can subsequently be lifted off the mesh, when the film will break around the grid leaving a continuous film over the whole grid area.

for examination or to pick up small fragments from a freshly cleaved surface. If a powder is to be mounted in this way then the 'etching' step should be replaced by dispersing the powder in a liquid and drying this down onto a glass slide, which is then coated with carbon. The 'replica' will generally lift off from a glass slide without a further etch.

Mounting and storing specimens

5.1. Grids

Specimen support grids are now almost universally 3.05 mm in diameter, except for a few high resolution stages and some very old instruments. They are available in a vast range of materials and designs. One catalogue lists 86 types in a total of 10 materials. The reason for this proliferation is to enable one to control the following:

(a) the amount of support the specimen needs (unsupported areas range from $20\,\mu m$ to 1 mm in extent);

(b) the material of the grid, so that it neither interferes with X-ray analysis nor reacts with the specimen;

(c) the labelling of specific regions of a specimen (many grids have identification marks for relocation of interesting fields).

The cheapest most widely used supports are copper grids at a spacing of 100 bars in^{-1}. Most grids have a shiny side and a dull side. Opinions differ as to the best side on which to mount the specimen but if a consistent practice is adopted it is always known which way up in the microscope the specimen was mounted.

5.2. Disc specimens

The advantage of preparing a 3 mm disc specimen is that it will need no support in the microscope. However, there is an exception to this rule. Many brittle or porous materials are liable to break up on handling and such materials can be supported by sticking them to a copper grid with a single large hole in the centre. This provides a ductile support to be handled by the tweezers without obscuring any of the thinned areas (with luck!).

5.3. Thinned foils

All foils except 3 mm disc specimens will need support in the microscope. In order to maximize the area available for viewing a grid with the widest spacing consistent with its supporting the specimen should be chosen. This could well be a 50 mesh grid. If a 400 or 700 mesh grid has to be used more than three-quarters of the specimen will not be visible. However, it should be remembered when examining a non-conducting or thermally sensitive specimen that the grid also acts as an electrical and thermal conductor and there would then be a reason for using a fine spacing.

Attaching the foil to the grid may be a problem. Some foils seem to stick remarkably well with no assistance. However, many foils will need to be secured lightly on the grid. Although folding grids, in which the foil can be trapped between two meshes, are available these are not recommended for 'materials' specimens since the specimen is nearly always damaged as the grid is folded. It is better in almost every way to use a sharpened toothpick to apply a very small drop of lacquer (e.g. Lacomit) to one or two corners of the specimen and to attach it to the grid just at these points. It is often easier to do this using a bench magnifier since the drop of lacquer should be of submillimetric dimensions.

5.4. Support films

Any specimen which cannot be supported by a grid alone requires a thin electron-transparent support film to be mounted on the grid first. Powders and fine particulate specimens are the most common examples. Many types of support films are in use but it is only necessary to describe three: the Formvar (polyvinyl formal) film which is suitable for fairly undemanding work, the carbon film when greater stability is required, and the perforated carbon film for high resolution work where the presence of any support films is to be avoided.

A Formvar film is very easily and quickly made. One drop of a solution of 0.5 per cent Formvar in chloroform is placed on a glass slide. A glass rod is used to wipe this drop uniformly over the whole slide. The resultant film will float off on to the surface of water as the slide is slid obliquely in (Fig. 24(a)). The easiest way to transfer the film onto 3 mm grids is to put several grids on a folded piece of metal gauze, submerge this in the dish and then bring the whole ensemble up beneath the floating film (Figure 24(b)). The continuous film will break around the edge of each grid as it is then lifted from the gauze, leaving each grid coated with Formvar. The stability of such films to the electron beam can be improved by coating them with a very thin layer of carbon. About 3 nm (scarcely visible on the porcelain dish (see Chapter 4) is adequate.

A pure carbon film is more stable than a coated Formvar film. This can be made by depositing carbon (again, as in Chapter 4) onto either freshly cleaved mica or a glass slide wiped with diluted Teepol and then wiped dry. In both cases the film, which can be of any desired thickness, should lift off very easily as the slide is immersed into water (Fig. 24(a)). The same technique as described in the previous paragraph can be used to transfer the film onto 3 mm grids. If the film is very thick it may need to be broken from around each grid with a needle.

The third type of support film is a perforated carbon film which can be used to support fine particles, some of which are likely to overhang a hole, or to carry a second carbon film which is too thin to be self-supporting. The manufacturing process combines both the previous techniques: to a 2 per cent solution of Formvar in chloroform is added 1–2 per cent of glycerin (the greater is the amount of glycerin, the larger is the number and size of holes). This mixture is agitated in an ultrasonic bath until the glycerin is well dispersed (emulsified) and then a glass slide is coated with one drop, as previously described for Formvar films. The glycerin is now dissolved out in methanol by immersing the whole slide in methanol for at least an hour. The slide is then allowed to dry and is carbon coated as described above for plain carbon films. The Formvar is then dissolved off in chloroform, either by immersing the whole slide in chloroform or by mounting the film on a grid which is then washed with chloroform as illustrated in Fig. 25. There are many ways of making a perforated carbon film, and a commonly used alternative procedure is to mount the Formvar films on grids before dissolving out the glycerin, carbon coating, and removing the Formvar.

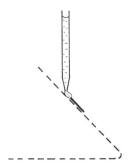

Fig. 25. Washing a grid or replica. The grid is placed on a piece of wire mesh folded so as to hold it at about 45°. A fine pipette is used to drop solvent gently above the grid, over which it will then wash.

Once mounted on a grid all the above films should be fairly stable. The coated grid can be immersed in further baths of liquid, for example to pick up a second film. Fine particulate material is generally mounted on a support film by dispersing it ultrasonically in an appropriate liquid (the choice depending on the wetting behaviour), placing a single drop of the specimen on a coated grid, and allowing it to dry. The key point is to ensure that the dilution of the suspension is correct; if it is too concentrated the particles will be piled on top of each other, and if it is too dilute the particles will be difficult to find on the support film. The correct dilution can only be established by trial and error.

5.5. Specimen storage

A Petri dish is adequate for temporary storage, say between preparing the specimen and deciding whether it is worth keeping. A piece of filter paper in the bottom can be marked into areas with a pencil and several specimens can be stored. It is possible to buy an insert for 50 and 90 mm dishes which is divided into numbered squares separated by raised ridges designed to minimize the risk of grids being mixed up.

Some sort of grid box is needed for permanent storage. The problems to be avoided are electrostatic charging of the grid, specimen, or box and mechanical damage to the specimen while it is inserted or removed from the holder. Several designs of plastic box with slits for 50 or 100 grids are available commercially. Grids can also be stored individually in gelatine capsules (diameter 5 mm) since only the edges of the grid or disc then touch the capsule and the risk of damage is minimized. Individual capsules can be stuck to a piece of card for labelling and storage. Some workers believe that the gelatine capsules can contribute to specimen-borne contamination in the microscope but the present author has seen no evidence of this.

The golden rules concerning specimen storage are: LABEL IT and DO NOT LEAVE SPECIMENS IN THE MICROSCOPE ROOM (they will disappear).

Appendix: Materials and suppliers

Most electron microscopy laboratories obtain the majority of their supplies from a specialist company which stocks a wide range of both consumable items and specimen preparation equipment. A short list of such general suppliers and some of the manufacturers of the larger items, together with an alphabetical list of all the items mentioned in the text with a reference to their distributors, is given below. One advantage of the large specialist suppliers such as Agar Aids is that it is sometimes possible to obtain via them good second-hand equipment. This can keep down the coat of establishing a new technique.

General electron microscope suppliers

1. Agar Aids Ltd.
   66a Cambridge Road
   Stansted
   Essex CM24 8DA
   Tel: (0279) 813519.

2. Balzers High Vacuum Ltd.
   Northbridge Road
   Berkhamsted
   Herts HP4 1EN
   Tel: (04427) 2181

3. EMscope Laboratories Ltd.
   374 Wandsworth Road
   London SW8 4TE
   Tel: (01) 720 6150

4. Polaron Equipment Ltd.
   21 Greenhill Crescent
   Holywell Industrial Estate
   Watford
   Herts WD1 8XG
   Tel: (0923) 37144

Suppliers of specific equipment

5. Edwards High Vacuum
   Manor Royal
   Crawley
   W. Sussex RH10 2LW

6. Gatan Inc.
   780 Commonwealth Drive
   Warrendale
   Pittsburgh
   PA 15086
   U.S.A.

7. IonTech Ltd.
   2 Park Street
   Teddington
   Middlesex TW11 0LT

8. Laser Technology Inc.
   10624 Ventura Boulevard
   N. Hollywood
   CA 91604
   U.S.A.

9. Logitech Ltd.
   Lomond Estate
   Alexandria
   Dunbartonshire G83 0TL

10. Materials Science (NW) Ltd.
    55 Cocker Street   Beverley Avenue
    Blackpool   Poulton-le-Fylde
    Lancs.   Lancs.

11. Metallurgical Services Ltd.
    Reliant Works
    Betchworth
    Surrey

12. Metals Research Ltd.
    Rustat Road
    Cambridge

13. Nanotech Ltd.
    Sedgley Park Trading Estate
    Prestwich
    Manchester M25 8WD

14. South Bay Technology Inc.
    4900 Santa Anita Ave
    El Monte
    CA 91731
    U.S.A.

15. Struers via
    Vickers Instruments Ltd.
    Haxby Road
    York YO3 7SD

16. Kerry Ultrasonics Ltd.
    Wilbury Way
    Hitchin
    Hertfordshire

Individual items

The numbers after each item refer to the suppliers listed above.

Acid saw 14
Adhesives 1, 2, 3, 4, 9
Aluminium wire 1, 2

Carbon rods 1, 2, 3, 4, 5, 13
Cellulose acetate 1, 2, 3, 4
Coating units 1, 2, 4, 5, 13
Cut-off machine 7, 10, 12

Disc cutter 7
Disc punch 10

Electropolishers 1, 10, 14, 15

Filter paper 1, 3
Formvar 1, 2, 3

Gelatin capsules 1, 2, 3
Gold wire 1, 3, 4
Grids 1, 2, 3, 4
Grid storage boxes 1, 2, 3, 4

Ion beam thinning equipment 1, 5, 6, 7

Lacomit 1, 2, 3, 4
Lacquer 1, 2, 3, 4
Lapping machines 1, 7, 9

Mechanical polishing jig 10
Mica 1, 2, 3, 4
MIcrorapid 7

Petri dishes 1, 3, 4
Polishing jig 10
Punch 10

Slitting wheel 11, 12
Spark machines 10, 12
Storage boxes 1, 2, 3, 4

Tenupol 1, 15
Tweezers 1, 2, 3, 4

Ultrasonic drill 16

Vacuum coating units 1, 2, 4, 5, 13

Wax 1, 2, 3, 4
Wire saws 8, 14

References

This is not intended to be an exhaustive bibliography, but merely to suggest a few sources which contain more detail than it has been possible to cover in this short text and to enable one or two specific topics which are not of wide applicability to be followed up.

Aitken, D., Tyler, S.K., and Goodhew, P.J. (1978). *J. Phys. E* **11**, 511–513. [Single-sided thinning]

Alderson, R.H. (1975). *Design of the electron microscope laboratory*, Chapter 9. North Holland, Amsterdam [Safety and hazards].

Franks, J. (1978). Ion beam technology applied to electron microscopy. *Adv. Electron. electron Phys.* **47**, 1–50. [Ion beam thinning]

Goodhew, P.J. (1972). *Specimen preparation in materials science* North-Holland, Amsterdam. [General text]

Von Heimendahl, M. (1980). *Electron microscopy of materials*, Chapter 2, Academic Press, New York. [General text]

Reid, N. (1978). *Practical methods in electron microscopy*, Vol. 3, part 2, *Ultramicrotomy*. North-Holland, Amsterdam. [Ultramicrotomy]

Smithells, C.J. (ed.) (1976). *Metals reference book* (4th edn.), Vol. 1, pp. 315–369. [Etchants]

Thomas, G. and Goringe, M.J. (1979). *Transmission electron microscopy of materials*, Appendix A, pp. 329–339. John Wiley, New York. [General text]

Unwin, P.N.T. and Wilkins, M.A. (1969). Preparation of thin foils from regions near cracks in massive bodies. *J. Phys. E* **2**, 736. [Cracks]

Willison, N.M. and Rowe, A.J. (1979). *Practical methods in electron microscopy*, Vol. 8, *Replica, shadowing and freeze etching*. North-Holland, Amsterdam. [General text]

Index

acid saw, 6
aluminium, 8 (T), 19 (T), 27
analysis, 3, 33
anodizing, 27
automatic electropolisher, 16

beryllium, 8 (T), 19 (T)
bismuth, 19 (T)

cadmium, 19 (T)
carbon coating, 30
carbon support film, 34
ceramics, 23, 28 (T)
chemical polishing, 7, 8, 24
    solutions, 8 (T)
chromium, 19 (T)
cleavage, 25, 32
cobalt, 8 (T), 19 (T)
contamination, 3, 36
cooling, 21
copper, 8 (T), 19 (T)
cracks, 26, 28
cross-sections, 27
cutting, 5

damage, 5, 6, 10, 11
dimpling, 10
disc, 10, 16, 33
drill, 11

electrolytes, 12, 17, 19 (T)
electropolishing, 12
    solutions, 19 (T)
etching, 13, 30
extraction replicas, 29, 30

foils, 33
Formvar films, 34
fracture surfaces, 26

gallium, 8 (T)
gelatin capsule, 35
germanium, 8 (T), 19 (T)
glass, 8 (T), 24

glycerin, 13, 34
gold, 20
grid, 33
    box, 35
grinding, 6
    jig, 7

hazards, 9 (T), 14
holey carbon film, 34
hollow anode gun, 23

ideal specimen, 2 (T), 4
indium, 19 (T)
initial preparation, 5 (T)
ion thinning, 10, 21, 27, 28 (T)
iron, 8 (T), 19 (T)

jet polisher, 16, 17

lacquering, 15, 26, 27, 34
lapping, 6, 7

magnesium, 8 (T), 19 (T), 24
mechanical polishing, 6, 7, 25, 26
mercury, 19 (T)
milling, 6
molybdenum, 19 (T)

nickel, 19 (T)
niobium, 8 (T), 20 (T), 27

one-sided thinning, 26

palladium, 20 (T)
platinum, 20 (T)
polymers, 28 (T)
powders, 31, 35
power supply, 14
punch, 10

recipes, 8, 17, 19 (T)
replicas, 29
rhenium, 20 (T)

saddle field gun, 23
safety, 9, 14
semiconductors, 8 (T), 24
shadowing, 29
silicon, 8 (T), 24
silver, 8 (T), 21
slitting wheel, 5
spark machine, 5, 11
sputtering, 21
    yield, 21, 22
stability, 3
stirring, 15, 16, 27
storage, 33
suppliers, 37
support grid, 3
    film, 34

tantalum 9 (T), 20 (T)
Tenupol, 16, 17

thickness, 3
tin, 20 (T)
titanium, 9 (T), 20 (T), 27
tungsten, 20 (T)
two-phase materials, 24, 28 (T), 30
two-stage replica, 29
typical specimen, 4

ultramicrotome, 25
uranium, 9 (T), 20 (T)

vanadium, 9 (T), 20 (T)

washing, 15, 21
window technique, 14, 24, 26
wire saw, 5, 6

zinc, 9 (T), 20 (T)
zirconium, 9 (T), 20 (T)